1 MONTH OF
FREE
READING

at

www.ForgottenBooks.com

By purchasing this book you are eligible for one month membership to ForgottenBooks.com, giving you unlimited access to our entire collection of over 1,000,000 titles via our web site and mobile apps.

To claim your free month visit: www.forgottenbooks.com/free917408

ISBN 978-0-265-97073-7
PIBN 10917408

This book is a reproduction of an important historical work. Forgotten Books uses state-of-the-art technology to digitally reconstruct the work, preserving the original format whilst repairing imperfections present in the aged copy. In rare cases, an imperfection in the original, such as a blemish or missing page, may be replicated in our edition. We do, however, repair the vast majority of imperfections successfully; any imperfections that remain are intentionally left to preserve the state of such historical works.

PARIS INTERNATIONAL EXHIBITION, 1900.

ECONOMIC MINERALS

OF

CANADA

BY

G. M. DAWSON, C.M.G., LL.D., F.R.S.

Director Geological Survey of Canada.

PRINTED BY DIRECTION OF

THE CANADIAN COMMISSION FOR THE EXHIBITION,

1900.

Economic Minerals of Canada

BY

GEORGE M. DAWSON, C. M. G., LL. D., F. R. S.

Director of the Geological Survey of Canada. *

The Dominion of Canada, comprising the northern half of a continent and exhibiting in different parts of its extent the most varied geological conditions, naturally affords many different minerals of economic value. Some of these have long been worked to a certain extent, but in late years a greatly increased development has occurred. This is best illustrated by the fact that in 1886 the value of the minerals produced in Canada amounted to $2.23 per head of the population, while at the present time it is about $8.90. In other words, mining is rapidly becoming one of the principal industries of the country.

The remarkable increase just alluded to, however, depends very largely upon the gold output, and particularly upon the great amount of alluvial gold that has in late years been drawn from the Klondike division in the Yukon District. Gold mining and especially the working of rich alluvial gold deposits is, from its nature, an industry that may be successfully carried on in tracts very remote from ordinary means of communication. Its prosecution attracts population and leads to permanent settlement, affording a means of opening up new regions to possession and agriculture ; but more general profit to the community undoubtedly results from the systematic

---

* In the preparation of this general sketch, for use in connection with the Paris Exhibition, the author has had the co-operation of several members of the staff, more particularly that of Messrs. Ells, Ingall, Denis and McLeish.

working of less intrinsically valuable minerals requiring for their proper utilization a greater amount of labour and skill. Mining industries of the last-mentioned class can scarcely be undertaken successfully elsewhere than in well inhabited districts or at points on the coast to which free access may be obtained by sea. Such industries, therefore, in a new country, extend with the spread of settlement and occupation of the land. They are of slower growth, but more permanent, and in all parts of Canada where railways, roads and water routes have opened up, industries of this class are now being rapidly establi ed.

It is intended here to give briefly the more important facts in this development, pointing out where ores and minerals of the various kinds are being produced, and in what directions activity of the kind is extending, with a few notes on the history and characters of the mineral industries. Particulars respecting the representative specimens comprised in the Canadian Exhibit at Paris, must be sought in the special Catalogue of this exhibit separately printed; while much detail in regard to current mining operations in the several provinces will be found in the reports of the provincial mining departments, as well as in the annual report of the Section of Mineral Statistics and Mines and in the Summary Report of the Geological Survey for 1899, for which enquiry may be made.

Some facts may in the first place be given in regard to the distribution of the more important economic minerals of Canada as a whole, coupled with a statement of the output of each of these for the last calendar year.

Coal, is abundant and extensively worked on both the Atlantic and Pacific coasts, its occurrence facilitating over-sea trade and local traffic on both oceans. The more important mines are situated in the provinces of Nova Scotia and British Columbia. New Brunswick produces small quantities of coal for local use, and coals and lignite-coals are largely developed in the Northwest Territories, east of the Rocky Mountains. There is no available coal in Ontario or Quebec, but in the first-named province the production of petroleum and natural gas to some extent takes the place of coal.

Iron, is found in important quantities in every province and probably in every district of Canada, but iron ores are being smelted only in Nova Scotia, Quebec and Ontario, and that on a limited scale at the present time, although important developments in this connection are now in progress.

Gold, is worked in the Yukon District, British Columbia, Nova Scotia, Ontario, Quebec and in certain rivers along the eastern base of the Rocky Mountains in Alberta and Athabasca, the value of the output being in the above order. In British Columbia the gold is derived both from placer deposits and from lode mining. In Ontario and Nova Scotia it results entirely from the latter kind of mining and elsewhere entirely from placer mines.

Silver, is to be credited almost entirely to the mines of British Columbia, where the working of argentiferous galena (or "silver-lead") ores is an important and increasing industry, but where it is also found in other associations. In Ontario, rich silver ores proper occur near the west end of Lake Superior and from Quebec a certain amount of silver is exported in association with ores of copper.

Copper, is produced by British Columbia, Ontario and Quebec, in the order named, largely in conjunction with gold and nickel, but high prices for the metal are now stimulating the development of copper ores proper.

Lead, in Canada, is again almost entirely derived from the mines of British Columbia, although deposits of galena occur also in other provinces.

Zinc, is also widely distributed and is often associated with other ores, but so far the small amounts shipped as zinc ores have been from Ontario and Quebec.

Nickel, is an important product of Canada, the only mines in operation being those of the Sudbury district in Ontario.

Manganese, in the form of its oxides, is raised in

variable, but not large quantities in Nova Scotia and New Brunswick.

Mercury, has been produced in small amount in British Columbia.

Platinum, occurs in connection with placer deposits of gold in British Columbia and a considerable amount of this metal is also contained in the nickel-copper ores of Sudbury, Ont.

Arsenic, is obtained in connection with the arsenical gold ores of Ontario.

Chromite or chromic iron-ore, is mined to a limited but increasing amount in Quebec.

Asbestus, in very important quantities, is also derived from Quebec.

Graphite or plumbago, is chiefly derived from the crystalline rock-series of Quebec and Ontario, but the product has not yet realized its full importance.

Gypsum, is extensively produced in and exported from Nova Scotia and New Brunswick. It is mined also in Ontario and known to occur in other provinces and districts.

Mica, is derived from Quebec, Ontario and (so far in small amount) from British Columbia. Its importance is rapidly increasing in connection with its use for electrical machinery.

Phosphate or apatite, of remarkable purity, is still produced in limited amount in Quebec and Ontario, although low prices in late years have greatly reduced the output.

Salt, is obtained in Ontario in quantities regulated only by the requirements of the market, while a certain small output is also to be credited to New Brunswick, and salt works have in previous years been in operation in Manitoba.

The output of *Petroleum* and *Natural Gas,* has already been alluded to. Other minerals appearing in the annual returns are : *Felspar, Fire-clay, Grindstones, Limestone,* (for flux) *Mineral pigments, Moulding sand, Pyrites* and

Soapstone. Structural materials and clay products show an aggregate estimated value in 1899, of nearly $5,600,000, including *Cement, Flagstones, Granite, Slate, Terra-cotta, Pottery,* etc., besides building materials proper, such as *Building-stone, Bricks, Lime,* etc.

Such is a general enumeration of the mineral products of Canada that have already become the basis of industries of greater or less magnitude. The subjoined table shows, in condensed form, the approximate amount and value of each of these products or classes of products in the calendar year 1899, together with the totals for each year from 1886 (the first for which complete returns are available) to date. In 1899 the minerals that may be classed as metallic, being utilized in that form, are valued at $28,833,717, the non-metallic minerals and structural materials at $18,141,795.

PRODUCTION OF MINERALS IN CANADA IN 1899.

PRODUCT.		QUANTITY.	VALUE.
			$
Copper (fine in ore, &c.) ....Lbs.		15,078,475	2,655,319
Gold, Yukon District...$16,000,000			
" all other...... 5,049,730			
		....	21,049,730
Iron ore............................. Tons.		77,158	248,372
Lead (fine, in ore, &c.)......... Lbs.		21,862,436	977,250
Nickel (fine, in ore, &c.)....... "		5,744,000	2,067,840
Platinum (partial return)... Oz.		55	835
Silver (fine, in ore, &c.)...... Lbs.		3,078,837	1,834,371
Arsenic . ... "		114,637	4,872
Asbestus and asbestic. .... Tons.		25,285	483,299
Chromite............................ "		1,980	23,760
Coal.................. .... "		4,565,993	9,040,058
Coke...... .... "		100,820	350,022
Felspar............................ "		3,000	6,000
Fire-clay......... .... "		599	1,295
Graphite....................... "		1,220	16,179
Grindstones................... "		4,511	43,265
Gypsum............................ "		244,566	257,329
Limestone (for flux)............ "		53,202	45,662
Manganese ore................... "		308	3,960
Mica............		....	163,000

PRODUCTION OF MINERALS IN CANADA IN 1899.
Continued.

PRODUCT.	QUANTITY.	VALUE.
Mineral pigments—		
Baryta.................... Tons.	720	4,402
Ochres............. .... "	3,919	19,900
Mineral water............................	....	100,000
Moulding sand............ .... Tons.	13,724	27,430
Natural gas...............................	....	387,271
Petroleum....... .... Bbls.	808,570	1,202,020
Phosphate (apatite)............ Tons.	3,000	18,000
Pyrites................. .... "	27,687	110,748
Salt................................... "	57,095	234,520
Soapstone............. .... "	450	1,960
Cement (natural rock)....... Bbls.	131,387	119,508
" (Portland)...... ...... "	255.366	513,983
Flagstones................................	....	7,600
Granite...................................	....	90,542
Pottery...................................	....	200,000
Sewer pipe...............................	....	161,546
Slate......................................	....	33,406
Terra-cotta...............................	....	220,258
Building material including bricks,	....	
building stone, lime, sands and	....	
gravels, and tiles...................	....	4,250,000
Estimated value of mineral pro-	....	
ducts not returned.....	....	300,000
Total, 1899......	....	47,275,512
1898, Total	....	38,661,010
1897 "	....	28,661,430
1896 "	....	22,584,513
1895 "	....	20,639,964
1894	....	19,931,158
1893	....	20,035,082
1892	....	16,628,417
1891	....	18,976,616
1890	....	16,763,353
1889	....	14,013,913
1888	....	12,518,894
1887	....	11,321,331
1886	....	10,221,255

The ton used is that of 2,000 lbs.

It will now be in order to refer separately to each of the provinces and the more important districts beyond the lines of the recognized provinces, beginning on the east and ending to the west in British Columbia and the Yukon District.

NOVA SCOTIA,

Although the total area of Nova Scotia does not much exceed 20,000 square miles, it offers a remarkable diversity of geological conditions and mineral resources. The principal minerals now worked are coal, iron and gold, with gypsum and various materials of construction. Besides these there are copper and lead deposits which have not yet become the basis of continuous industries, as well as manganese, antimony and other minerals which have been worked irregularly or of which the existence is known.

Nova Scotia was known from very early days to possess important mineral deposits, but these only began to attract attention in the first part of this century, and their exploitation on any considerable scale practically dates from 1830, when the first deep shaft of the General Mining Association was sunk on a coal-seam. Since then the development and working of some of the mineral deposits have been pushed actively, while others, although apparently promising yet remain undeveloped.

Mineral production in Nova Scotia in 1899 :—

Coal, tons	3,148,822	$4,920,035
Gold, ounces	29,879	617,604
Iron ore, tons	24,232	72,696
Gypsum, tons	126,754.	102,055

Coal.—The Carboniferous formation occupies a large portion of Nova Scotia, covering over half the area of Cape Breton Island as well as a large part of Cumberland, Pictou and Hants counties. Coal constitutes the main mineral product of the province.

The first mention of the existence of coal occurred in 1672, when, in a geographical and historical notice of the

American coasts, Nicholas Denys says that on some mineral concessions on the Island of Cape Breton there were mines of a coal equal in quality to the Scotch. It was not until 1784, however, that underground work was attempted. Before that time the yearly quantity of coal mined from the outcrops never exceeded 3,000 chaldrons. In 1827 the Sydney mines became the property of the General Mining Association, and this company had, further, practically the undisputed possession of all the coal-fields of Cape Breton until the year 1857, when they surrendered their claims, except within certain well-defined limits. After that date more liberal terms were offered to miners and several leases were taken up, and work begun on them.

The coals mined in Nova Scotia are of the bituminous class and are generally well adapted to the manufacture of gas and coke. Excellent steam coal is also produced.

The following table shows the disposal of the coal mined in Nova Scotia for the year 1899 :—

	Tons.
Nova Scotia	841,169
New Brunswick	370,485
P. E. Island	76,622
Quebec	1,214,410
Newfoundland	120,163
West Indies	6,769
United States	320,449
	2,950,067

The shipping facilities are all that can be desired. In many cases the coal is loaded in ships directly from the mines and most of the mines also have good railway connections.

The home market for Nova Scotia coal is as yet limited ; the greater part of Ontario, and even some parts of Quebec are supplied from the United States fields which are nearer, or which afford anthracite, not found in Nova Scotia. The iron industry of the province seems, however, to have now entered an era of development which will greatly increase the prosperity of the coal mines.

A few notes may be given respecting the principal coal-fields, as follows :—

The Sydney Coal Field is situated in the north-east corner of Cape Breton county, and takes in a small portion of Victoria county. It occupies a land area of 200 square miles, being about 32 miles long by 6 wide, and is limited on three sides by the Atlantic Ocean. The conditions for extraction and shipment are very favorable ; there is a remarkable absence of faults and the coast affords good natural harbours. A greater part of the field extends under the sea. Within this area the existence of 9 different seams has been recognized, ranging in thickness from 3 to 12 feet ; and these dip at low angles, 5° to 12°, seaward. On these are now working eleven collieries, the output of some amounting to as much as 350,000 tons a year. The total production of the field for the year 1899 was over 1,700,000 tons. As a rule the mines of this field are very completely equipped. Under-ground haulage is effected in most of them by endless rope system, and the exploitation is conducted according to modern and systematic methods.

The Cumberland Field is the most westerly of the coal districts of Nova Scotia, a part of it being adjacent to Chignecto Bay. In this field there are two coal-producing areas. One situated near the coast, which may be called the Joggins coal-basin, and the other situated about 15 miles to the east of the first, at Springhill.

In the Joggins area seven seams are known, ranging in thickness from 2½ feet to 9½ feet.

The Springhill area contains eight seams with thicknesses of 2½ to 13 feet. In both fields are well equipped collieries worked by slopes driven on the seams. The total production of the Cumberland field for the year 1899 amounted to nearly 450,000 tons.

The Pictou Coal Field, situated almost in the centre of Pictou county, has an area of productive measures of about 25 square miles. It is eleven miles long, with a maximum breadth of 3 miles. This field is therefore small, but some of the seams are of great size, one being 38 feet thick. The district is of very intricate structure, being cut up by numerous faults of varying magnitude, and the productive measures are almost completely surrounded by a girdle of faults.

The field was opened in 1798, but the first systematic work was contemporary with the development of the Cape Breton field in 1827, when both became the property of the General Mining Association. It is now the scene of operation of several collieries, which at present produce at the rate of about 450,000 tons annually.

The Inverness and Richmond Coal Fields are not so important as those above mentioned, and their production is small at present but likely to be increased.

Iron.—In the year 1604, Lieutenant General des Monts, while surveying a part of the coast, noted layers of magnetic iron sand on the beach of St. Mary Bay, and also veins of iron ore in the trap rocks of Digby county; but the first attempt to utilize iron ore deposits in Nova Scotia was not made until the first decade of the present century; when a Catalan forge was built at Nictaux, where a few tons of bar-iron were produced. In 1825 a comparatively important company, with a capital of £10,000, established smelting works on the east bank of the Moose River, which, although only in operation a short time, produced an excellent charcoal iron. An attempt to smelt the iron ore of Pictou county was made in 1828 by the General Mining Association. A small furnace was erected near the Albion mines, but seems to have given unsatisfactory results.

The next attempt was made at Londonderry, Colchester county, on ores from that locality, and this proved more successful. The Acadia Iron Works, after employing a Catalan forge for a few years, erected a furnace in 1853 and ran until 1874, producing in all some 45,000 tons of pig-iron. In 1873 the Steel Company of Canada acquired these works and extended them considerably, spending two and a half million dollars in the building of blast-furnaces, mills, etc., and in securing iron ore deposits and a colliery. The company did not meet with the success expected, and was reorganized in 1887 under the name of the Londonderry Iron Co. This company has extensive works, consisting of a blast-furnace, coke-ovens, rolling mills, foundry, rail connection, etc. It also owns four iron mines and the Maccan coal areas.

Besides the above the companies at present existent in Nova Scotia for the production of iron and steel are : (1) TheNova Scotia Steel Company. The capital of this company is $5,000,000. It has a furnace at Ferrona, which uses a mixture of foreign and native ores. The steel-works are at New Glasgow. (2) The Pictou Charcoal Iron Company, which was formed to manufacture charcoal pig-iron on the East River, Pictou county.

The latest and most important development, however, of the iron and steel industry in Nova Scotia, is the formation of the Dominion Iron and Steel Co. of Sydney, Cape Breton, with a capital of $20,000,000. This company is now erecting a very extensive plant at Sydney Harbour.

The smelting facilities afforded by this province, which possesses valuable coal fields and deposits of iron ore, beside being within easy reach of foreign ores, suggest the utilization of its iron product in manufactures and industries on a much larger scale than has hitherto occurred.

All the different varieties of iron ores are met with in Nova Scotia, from hæmatite and magnetite to bog ores. Many of the deposits are as yet undeveloped or have not had enough work expended on them to enable their value to be estimated, others are known to be capable of affording great supplies of ore.

In Cape Breton, deposits of red hæmatite are known at Big Pond, Red Islands, Loran, and several places on the Bras d'Or lakes. None of these have as yet been systematically mined but some of them may prove to be of considerable extent. At Whycocomagh, red hæmatite and magnetite are found in close proximity. Nine beds are reported to have been discovered and partly tested, varying in thickness from 3 to 9 feet. A large bed, near

Gillies Lake, has been traced for over 2 miles and has a thickness of from 4 to 13 feet.

In Guysborough county, several important deposits have been opened, as at Erinville, where exploratory work is said to have exposed a width of 60 feet of specular ore. In Antigonish county lenticular masses of spathic ore are found at Polson Lake, etc. Beds of hæmatite are known at Arisaig. In Pictou county important deposits of iron ores occur, including specular ore, bedded red hæmatites, limonite and spathic iron. Some of these have already been worked to a considerable extent, particularly the remarkably pure limonites of the East River valley. In Colchester county, a belt of strata along the southern flank of the Cobequid Mountains is found to carry important quantities of carbonates and oxides of iron. Large bodies of limonite have been found here, and on these extensive mining operations have been conducted. In Annapolis and Digby counties, in the western part of the province, magnetites and hæmatites are abundantly developed. The most important deposits are bedded ores of Devonian or Silurian age, and often fossiliferous. Torbrook, Nictaux and Clementport are the best known localities and considerable quantities of ore have been mined.

Gold.—The first discovery of gold-bearing quartz in Nova Scotia is said to have been made by Lieutenant C. L'Estrange, R.A., in 1858, while on a hunting expedition in the woods at Tangier. No importance, however, seems to have been attached to this find, and the discovery which attracted attention to these gold fields was that of John Pulsiver, a farmer, in 1860. In 1862 the Government appointed a Gold Commissioner, and the gold mining industry of the province may be said to date from that year.

In Nova Scotia the gold-bearing rocks form a belt, varying in width from 10 to 70 miles, and extending some 260 miles in length along the Atlantic coast, covering an area of about 5000 square miles. These rocks, which consist of quartzites and slates, have been folded into a series of synclines and anticlines nearly parallel to the coast-line, and with axes about three miles apart. The recent studies of the Geological Survey of Canada shew that, along the crests, at intervals varying from ten to twenty miles, the anticlines have a decided pitch in opposite directions, giving rise to long elliptical domes. On these domes the strata have loosened and opened up, affording room for the introduction of auriferous quartz in a succession of superimposed saddle-shaped veins, like those of Bendigo in Australia. Each one of these domes is thus the centre of a system of gold-bearing veins, and some fifty such centres are now exploited. So far, however, gold mining in Nova Scotia has been limited to veins that outcrop at the surface, and workings have not reached a greater depth than 700 feet, whereas the Bendigo reefs have been mined to a depth of 4000 feet. The work now being done at the Dufferin mine in Halifax county, if successful, will open a new era in gold mining in Nova Scotia, by inaugurating a more comprehensive method of working systems of veins collectively.

The quartz is free-milling and the quality of the gold is very high. More attention has of late been given to the economic working of the lower grades of quartz, and large quantities containing only 1 dwt. 18 grs. to the ton have been successfully dealt with. The average yield per ton from some districts is, however, as high as one and a half ounces ; while other districts, with small production of a few hundred tons of quartz, have averaged much higher, one of these in 1899 having run up to 3 oz. 16 dwt. per ton, on a production of 131 tons.

The aggregate yield for the province is remarkably constant, having been, during the past ten years, as follows :

1890.	$ 474,990
1891.	451,503
1892.	389,965
1893.	381,095
1894.	389,338
1895.	453,119
1896.	493,568

1897.	562,165
1898.	538,590
1899.	617,604

Copper.—Very early explorers of Nova Scotia mention the occurrence of native copper in association with the trap rocks of the Bay of Fundy, but it is only within the past twenty-five or thirty years that the copper deposits of the province have attracted the notice of miners. Amongst these deposits are those of Coxheath, near Sydney, Cape Breton Island. The ores occur as copper-pyrites in pre-Cambrian felsites, and a good deal of development work has been done on them. They have been favourably reported on but have not been worked as yet to any considerable extent.

Copper ores are known in many other places both in Cape Breton and in Nova Scotia proper, but no regular output has yet occurred. A copper smelting plant under construction at Pictou may, however, result in the actual development of some of these deposits.

Ores of manganese occur in workable quantities in association with some of the limestones of the Carboniferous system, being found in the counties of Colchester, Pictou, Cumberland and Hants. The ores consist of pyrolusite and manganite and some of them are very pure. The deposits on which most work has been done are those of Teny Cape in Hants county. Manganese oxides also occur in association with some of the limonite iron ores and with bog-iron ore.

Stibnite, or sulphide of antimony, is known in several places in Nova Scotia, and a deposit was somewhat extensively worked at Rawdon, Hants county, for antimony. This was subsequently abandoned, but at a later date it was discovered that this ore of antimony was auriferous, leading to a resumption of the work, gold being the principal value and antimony a by-product.

Deposits of gypsum in Nova Scotia are on a very large scale. The beds are frequently traceable for miles and sometimes present faces fifty feet in thickness. These beds are included in the Carboniferous system, often associated with limestone. The gypsum produced is of all degrees of purity but is often of very high grade and may be employed in the manufacture of the finest qualities of plaster of Paris. The greater part of the gypsum produced is exported in the crude state to the United States, particularly from Windsor in Hants county, where about 150,000 tons is shipped annually. The less pure gypsums are used extensively as a fertilizer under the name of " land plaster." The chief localities in which gypsum occurs in important quantities are Windsor, Cheverie, Maitland, Walton, Wallace, Antigonish, Judique and many places on Bras d'Or Lake and in Pictou county.

Sandstones suitable for building or for the manufacture of grindstones, scythe-stones, etc., are abundant in Nova Scotia; as well as granites, syenites and other stones well adapted for purposes of construction. Marble occurs in several parts of Cape Breton Island. Plumbago has been found in the form of plumbaginous shales in several places in Cape Breton, and these shales have already been worked in an experimental way. Some of the clay-slates of the Silurian formation have been successfully used for lining cupolas, furnaces, etc., and decomposed felsite found in the Coxheath Hills has been ascertained to possess refractory qualities that would fit it for the manufacture of fire-brick.

NEW BRUNSWICK.

In New Brunswick, at the present time, the principal mineral products utilized are gypsum, lime, coal and mineral water, stones for the manufacture of grindstone and for building purposes, and clays employed in making various clay products. Formerly iron, manganese and albertite were considered amongst the important minerals mined in this province, but for various reasons the production of these has fallen away, though in the case of manganese an important revival of attention has recently taken place. Mineral products, both metallic and non-metallic are, however, widely distributed throughout the province, and many of these are no doubt susceptible of successful and extensive development.

Following is the production of some of the chief minerals in New Brunswick, for which statistics are available in 1899 :—

		Quantity.	Value.
Coal,	tons.............	10,528	$ 15,792
Grindstones,	"	3,133	32,965
Gypsum,	"	116,792	151,296
Granite,	....		20,070
Mineral water,	....		15,000

Iron ores are found in Carleton county and are most readily accessible at Jacksontown near Woodstock. The Woodstock hæmatite deposits are associated with a series of slates, usually bluish or grayish in colour and highly calcareous, but when in connection with the iron ores becoming more or less reddish or greenish. These ores were first brought to public notice in 1836 by Dr. C. T. Jackson, of Boston, and were subsequently utilized in the blast-furnace operations begun at Woodstock in 1848 and carried on intermittently for 20 years thereafter.

The operations were subject to so many interruptions, however, that the whole time during which the principal furnace was in blast is said to have been only about eight years.

Other occurrences of iron ore have been noted at West Beach and Black River, on the Bay of Fundy, near St. John, and also in Charlotte county, near the village of Lepreau and on Deer Island.

Ores of antimony, consisting of stibnite or sulphide of antimony with occasional specimens of native antimony have long been known in the parish of Prince William, York county. The first company to undertake active operations was the Lake George Mining and Smelting Co. A considerable quantity of ore was raised and an extensive plant was erected. When in full operation these works yielded fifteen tons of metal every six weeks.

The product was partly exported in ingots and partly employed on the spot in the manufacture of babbit metal by admixture with lead, copper and tin. Owing to a decrease in the demand for the metal and other causes, though not to deterioration in either the quantity or quality of the ore, all work ceased about 1890 and has not since been resumed.

Ores of copper, including native copper, cuprite and copper sulphides have been found at various places throughout the province, but no successful attempts at mining have been made.

Deposits of nickeliferous pyrrhotite, near St. Stephen, have attracted some attention. They somewhat resemble those at Sudbury, Ont., though having a more variable and lower percentage of nickel. No successful exploitation of the deposit has yet occurred.

Of the several deposits of manganese found in New Brunswick that at Markhamville, Kings county, is the most interesting and important. The mines are situated near the head of the Hammond River, about forty miles north-east of the City of St. John. The first systematic operations for the extraction of the ore are said to have been undertaken by Colonel Markham, on behalf of the Victoria Manganese Co., about the year 1864, and since that time operations have been continued with greater or less regularity until recent years, the mines being finally closed down about 1893. At first only superficial deposits were worked, consisting of ore found in pockets in beds of clay, mingled more or less with gravel. Subsequently, operations were extended to the underlying limestones, but here also the distribution of the ore was found to be exceedingly irregular, leading to such fluctuations in the output that, eventually, search for new deposits proved profitless and mining was discontinued. The total production of these mines is reported to have been over 23,000 tons, largely of very high-grade ore.

Other deposits similar to those at Markhamville are found at Jordan Mountain, Kings county, at Shepody Mountain, Albert Co., and at Quacco Head, St. John county. Some work has been done at each of these deposits.

Another class of manganese ores occurring in New Brunswick is the bog-manganese or wad found in Dawson settlement, Albert county, and elsewhere. These, though they have hitherto been considered of little value, have recently been utilized in the manufacture of ferro-manganese and may yet prove to be a considerable mineral asset to the province.

Bituminous coal was one of the first minerals to attract attention in New Brunswick, and is said to have been mined in small quantities in the Grand Lake district as early as 1782.   The coal mines at Grand Lake are situated mainly in the vicinity of the Newcastle River, on the Salmon River, in Chipman, and about the lower part of Coal Creek ; the entire extent of the Newcastle basin being estimated at about 100 square miles.   The development of the mines has been very slow and throughout their history there has been almost a total lack of combined or persistent effort.   For many years the removal of the coal was effected in a most desultory way, each farmer upon whose land the seam was exposed devoting a portion of his winter leisure to getting out what was needed for his own use or occasionally hauling a load to Fredericton.

There is but one seam of coal, about 22 inches in thickness, in this district, but it occurs often so near the surface that it may be obtained by the simple process of stripping and quarrying.   The overlying drift varies in thickness from a few inches to thirty or forty feet.   The output of the Grand Lake coal mines has averaged about 6000 tons per annum since 1863.

It is possible that coal-seams of greater thickness may still be found in other parts of the province, and boring operations with the object of testing this are about to be undertaken on a considerable scale.

Albertite, although no longer mined, was a mineral formerly of great importance in New Brunswick.   It is a species of mineral pitch, black in colour and breaking with a conchoidal fracture.   The mineral was originally discovered in the year 1849, near the town of Hillsborough, the vein at the surface showing a thickness of sixteen feet. From 1863 to 1874 over 150,000 tons were mined. It was exported largely to the United States, where it was used partly as an enricher in the manufacture of gas and partly in the making of oil. The deposit finally became exhausted and no new veins of workable dimensions have yet been discovered.

The mining of gypsum is one of the most important mineral industries of the province, the value of the product in 1899 being over $150,000. The mineral is found in the rocks of the Carboniferous system, and is somewhat widely distributed. The most extensive deposits and those which have attracted the greatest attention are situated in Albert county, notably at Hillsborough. These have been mined for many years, regular shipments having been made since as early as 1854. The superior quality of the "plaster" made from these deposits has gained for them a wide reputation. In quarrying the gypsum, ordinary black powder is employed. The process of manufacturing is to dry the gypsum, then grind and calcine it. Shipments are made to all important points in Canada and to some points in the western United States. The works are operated by the Albert Manufacturing Co.

Graphite has been found in St. John county near the mouth of the St. John River. The limestones here carry the graphite more or less disseminated, but in places in sufficient bodies to admit of being worked. Although mining has been attempted at several times, operations have never been extensive and have been quite irregular.

Salt springs occur at several places in Kings county, notably at Sussex, where, since 1887, about 150 barrels of salt have been made per annum. Mineral waters are also found in the same vicinity, at Sussex and at Havelock. These waters are employed somewhat extensively as beverages.

Freestone, and stone suitable for grindstones, etc., are found in abundance throughout the rocks of the

Carboniferous system in New Brunswick. The grindstone industry formerly centered around the head of the Bay of Fundy, in the southern portions of Albert and Westmorland counties, at Grindstone Island, Mary Point, etc. Over 50,000 grindstones are said to have been made here in 1851. Grindstones and building stone are now quarried at Woodpoint quarry, near Sackville, and at Cobourg quarry, near Bay Verte and work has been done in the parish of Dorchester. The industry has also attained considerable importance in the north, about Newcastle, in Northumberland county, and Stonehaven and Clifton in the Bay of Chaleurs. From the French Fort quarry, near Newcastle, much sandstone of a very superior and durable quality has been taken. It has been used in the construction of the Langevin Block at Ottawa and in other works of importance. Some grades of it are admirably suited for the manufacture of stone for wood-pulp grinding.

The freestones of Clifton and Stonehaven are said to be less suited for building, although they have been used for that purpose. By far the larger part of the rock is used in the manufacture of grindstones, with pulp-stones, sythe-stones, etc., as subordinate products.

Granite from Hampstead, Queens county, known as Spoon Island granite, attracted early notice, although the quarrying industry has not become very extensive there. The red granites of St. George, Charlotte county, are better known, and the latter town has become the seat of somewhat important works. The stone has been used in many buildings, both public and private, and in bridge work. It is also excellently adapted to monumental work and a considerable industry is carried on in cutting and polishing monuments, columns, etc., by water-power.

Limestones are abundant throughout the province, but the remarkable purity of the deposits near St. John, with the facilities afforded for working them, have produced an important industry. Lime is sent to many adjacent ports. Clays suitable for the manufacture of bricks occur in many places, and bricks are manufactured both for use in the province and for export.

PRINCE EDWARD ISLAND.

This little province, consisting of the island of the same name in the Gulf of St. Lawrence, has practically no mineral industries. It is uniformly fertile and well peopled and has important fisheries. Red sandstones suitable for masonry may be quarried in some places, and clays occur for brick-making. It may be that coal-seams underlie the island or some part of it, but if so they appear to be at a depth too great for utilization at the present time.

QUEBEC.

The province of Quebec may be said to take a high place as a producer of economic minerals. In some lines, notably asbestus, graphite and apatite, it takes first rank in Canada, and the output from the asbestus mines has in late years largely controlled the world's market. In other lines such as gold, iron, copper and mica, the deposits are very extensive, and for some years mining has been carried on, although the output has been affected largely at times by the question of supply and demand through competition with other sources. Mines which are now closed owing to this factor, as in the case of apatite, will no doubt again become paying properties with changed conditions.

The following table gives, in summarized form, a statement of the mineral production of Quebec for the year 1899.

		QUANTITY.	VALUE.
			$
Copper	Lbs.	1,632,560	287,494
Gold	Oz.	238	4,916
Iron ore	Tons.	19,420	50,161
Silver	Oz.	40,231	23,970
Asbestus	Tons.	25,285	483,299
Chromite	"	2,010	21,842
Mica	"	571	136,863
Ochres	"	3,894	19,650
Pyrites	"	27,687	110,748
Cement	Bbls.	19,546	32,871
Slate	Tons.	2,664	30,406
Granite	"	9,895	62,062

The iron ore deposits of this province were among the first mineral products to be utilized. The ores are of various qualities, including magnetite, hæmatite and limonite or bog-iron ore. Of the former, large deposits are often titaniferous, and though attempts have been made from time to time to smelt these, notably along the lower St. Lawrence, the attempts have been attended with failure, owing to the refractory nature of the ore. While the magnetic ores have been worked at a number of points in the province, both in the Ottawa district and in the "Eastern Townships," the principal operations in iron smelting have been confined to the bog-iron ores of the St. Francis River, east of the St. Lawrence and to those found in the vicinity of the St. Maurice River to the west. At this place the importance of the ore-deposits was recognized by the early French settlers more than two centuries ago, and a report was made on them to the French government as early as 1681. Smelting operations were commenced in this district in 1733 and the industry then begun, has been carried on to the present time.

The early furnaces in this district were small, having a daily capacity of not more than five to six tons. These, some years ago, were replaced by a greatly enlarged plant, the furnaces having a daily capacity of some fifty tons of pig-iron. The product is of excellent quality and admirably adapted to the manufacture of car-wheels, for which purpose a large portion of the output has long been employed. The supply of ore is largely obtained by dredging operations carried on in Lac à la Tortue.

Along the Ottawa, extensive deposits of magnetic iron-ore are found near Hull, where smelting operations were carried on thirty years ago, and in the township of Bristol, about forty miles west of Ottawa city. Both these have been quite extensively worked in earlier years, and mining has again been resumed lately. All these ores appear to be associated with the crystalline rocks of the Archæan.

In the Eastern Townships, large bodies of magnetic ore, in places containing titanic acid, are found associated with the rocks of the Cambrian series and also with those supposed to belong to the Huronian period and forming the Sutton mountain chain.

Copper ores are somewhat widely disseminated in the Eastern Townships of Quebec. They occur principally in connection with the crystalline pre-Cambrian rocks in the form of sulphides. The widely distributed character of the ores may be gathered from the localities enumerated in the Geological Report for 1866, where a list of fifty-six places is given in which these minerals have been found in greater or less quantity. Several of these localities have been extensively worked in former years, but operations are at present chiefly confined to the mines at Capelton and places in this immediate vicinity where there is a large body of ore. The ore is chiefly shipped to the United States, where it is first employed in the manufacture of sulphuric acid, the copper contents, which average from three to four per cent., being subsequently refined, while there is also an appreciable amount of silver, stated to average about four ounces to the ton. At the Harvey Hill mines, also, very extensive operations were at one time carried out, and a large amount of copper has been extracted. At Acton, on the line of the Grand Trunk Railway, about sixty miles east of Montreal, important deposits of copper ore were mined about forty years ago. These occur in slates assigned to the Cambrian and cut by diorites.

The first reference to the copper ores of Quebec was made in the early reports of the Geological Survey for 1847-48. Mining operations commenced on the Acton deposit about the year 1858 and at the Harvey Hill mines about the same date, while the work on the Capelton mines was begun several years later. The history of these mining enterprises in Quebec will be found in the Report on the Mineral Resources of Quebec for 1888, by Dr. Ells of the Geological Survey.

The mining of asbestos in Quebec is a comparatively recent industry. At Black Lake and Thetford, about seventy miles north of the city of Sherbrooke, the large deposits of chrysotile, or serpentine-asbestos, which have since become so well known, were discovered in 1877-78. The principal deposits occur at two points, the most important being at Thetford, where the containing serpentine-rock forms a considerable area surrounded by Cambrian slates. The largest mines are located on a knoll about eighty feet above the railway track. The mineral occurs in a series of veins which range from small threads up to a thickness in places of about six inches, though those of the large size are comparatively rare. Veins of three and four inches of fine fibre were, in the first year of the working, quite plentiful; but as the mines increase in depth these appear to diminish somewhat in size. The veins reticulate through the rock in all directions. They are worked by open quarrying, the larger veins being readily broken out while the smaller material is carefully cobbed. The separation was formerly accomplished altogether by hand, but within the last few years machinery has been employed. A material named ''asbestic'' is now also manufactured near Danville, by crushing the serpentine rock containing more or less short asbestos fibre in association. This is employed as a fire-proof plaster for building purposes.

Openings for asbestus have been made at a number of other points on the serpentine belt in the Eastern Townships. Among these may be mentioned the mines at Broughton, at Coleraine, and those near Brompton Lake. At none of these places, however, is the development, in so far as yet ascertained, so extensive as at the mines at Thetford or Black Lake.

Intimately connected with the same wide belt of serpentine rocks, in Quebec, are the deposits of chromic iron which have recently come into prominence in the mineral output of the province. The mineral occurs in a series of irregular pockets in a serpentine which is practically barren of workable veins of asbestus in so far as yet ascertained. The principal deposits are found to the east and south of Black Lake station, on the line of the Quebec Central Railway, about sixty-five miles north of Sherbrooke. Deposits of this mineral were found as early as 1865 in the area to the west of Coleraine station. The mineral is now obtained at various points in the townships of Wolfestown, Ireland, Garthby and Coleraine, but the principal workings are at present situated to the south-east of Black Lake. The total amount raised for shipment from 1894 to 1898, in this district, was rather more than 10,000 long tons. The sesquioxide of chromium in the ore ranges from 40 to over 60 per cent., a fair proportion being classed as high-grade ore. As the extent of the serpentine belt is large in this district, the chances of finding workable deposits at many points are very good. Chromic iron has also been found in the extension of the same rock series as far as the Gaspé peninsula.

The gold mines of eastern Quebec have been known for many years. The mineral was first recognized in the Chaudière district about 1824, and a second and more important find was made in 1834. Mining was first commenced in 1847 and it has been prosecuted at intervals to the present time. The sources of the Chaudière gold have long been discussed, but no quartz veins or other fixed deposits of a payable character have yet been found. It is no doubt, however, derived from such deposits in the Cambrian or pre-Cambrian rocks of the vicinity. The mining so far done has been principally in the gravels of old river-channels tributary to the Chaudière River. Some of these have been found to have a depth of over 160 feet and to be overlain by glacial deposits.

The gold is generally coarse, and nuggets of a value of over $1000 have been obtained from the gravels. Difficulty is sometimes found in working these old channels, but some of them are without doubt very rich in gold.

Another area of gold mining is found in the south-eastern portion of the province on what is known as the Ditton River. Operations were carried on in this locality for some years, and a considerable amount of coarse gold was obtained. The work has been conducted in a desultory manner in recent years with varying success.

On the flanks of the ridge of crystalline rocks that

traverses the Eastern Townships some attention has been given to the prospecting of lodes. Near Dudswell, visible gold has been found in the quartz veins, as well as at a place called Handkerchief, near the northern portion of the township of Leeds. These discoveries are interesting, but have not yet been proved to be important.

Mica is mined principally in the region to the north of the Ottawa River and in the vicinity of the Lièvre and Gatineau rivers. The deposits have been somewhat extensively worked for the last ten years. The mica is frequently associated with deposits of apatite, when both these minerals are produced in economic quantity from the same mine. As a rule the mica occurs in irregular veins traversing pyroxene-rocks. The finest crystals are usually obtained from a matrix of pinkish or sometimes greyish calcite, in the pyroxene mass. The pyroxene occurs in bodies of apparently intrusive and much altered rock that cut the gneisses and limestones of Archæan age. The bulk of the mica is the variety known as " amber mica" or phlogopite, although muscovite is also found at some points. The latter is a constituent of pegmatite dikes and is not found in the pyroxene-rock. It is found in considerable quantity along the Lower St. Lawrence and has there been mined to some extent. Along the Gatineau, and in the area between this river and Lièvre River, the phlogopite deposits are numerous and valuable. Though the mining of mica is a comparatively new industry in this area, the output has already reached large dimensions. The deposits are, however, frequently very irregular and a large proportion of the output is not marketable. The quality is excellent and is well adapted for electrical purposes.

Closely associated with the mica are large deposits of apatite or phosphate of lime. This mineral occurs also in connection with the pyroxene-rocks of the district, and the workable mines are mostly confined to the district along the Lièvre and between this river and the Gatineau.

The discovery of apatite in the province of Quebec

was made about the year 1871, and mining was commenced shortly after. The industry gradually became
important and was successfully prosecuted for nearly
twenty years, or until the discovery of the Florida phos-
phates, which so lowered the price of the Quebec
mineral that mining became unprofitable. For the last
last six years no attempt has been made to work apatite
in Quebec except as a by-product in connection with
the mining of mica. The output of the mineral for 1889
was stated to be about 33,000 tons. The mineral is generally of high grade, and the quantity is practically inex-
haustible in the Lièvre district. It occurs principally
in the form of pockets and sometimes as veins, which
have in places been followed to a considerable depth,
some of the mines being worked to a depth of over
600 feet without appreciable diminution. From some
of the pockets more than a thousand tons of first
class mineral have been obtained by simple quarrying.
Lack of shipping facilities in some places is a serious
obstacle to profitable development.

The graphite or plumbago deposits on the north of the
Ottawa, from Grenville westward, were early recognized
and were referred to in the reports of the Geological Survey
for 1845-46. The excellence of the mineral was noted by
Dr. T. S. Hunt many years ago, and some attempts at mining were then made. The principal deposits, however,
are situated near Ottawa, in the township of Buckingham.
Here the graphite occurs generally disseminated through
a greyish gneiss, and in places veins of columnar graphite
of good size are also found. The quantity in this area is
apparently very great. Mills have been erected at the
three mines, and the mining is now being conducted in
a systematic manner with the promise of satisfactory
results. The prepared ore is equal to the best Ceylon
graphite, as shown by a series of analyses made some years
ago by Dr. Hoffmann, of the Geological Survey. Other
deposits of importance are found on the east side of the
Lièvre in the township of Lochaber. These have also
been to some extent developed, and work is again to be
shortly resumed in this locality.

Silver is found at a number of points in the province of Quebec, for the most part associated with galena. The metal also occurs in association with the copper ores of the Eastern Townships, as already noted, and has been found as far east as Gaspé Basin.

Among other minerals of more or less importance may be mentioned molybdenite, which has been found at a number of points, but of which the largest deposit so far recognized is on the upper Gatineau River, in the township of Egan. None of these deposits have yet, however, been worked. Ochres, extensively used for the manufacture of paints, are found in many places. The principal deposits are situated along the St. Lawrence in the vicinity of Three Rivers. Magnesite occurs in beds of large extent at several places in the Eastern Townships, notably in the townships of Bolton and Sutton, the percentage of carbonate of magnesia ranging from 59 to 83 per cent. Felspar, suitable for the manu-facture of pottery, is found at many points in the Lau-rentian area as an ingredient of pegmatite dikes, and has been quarried and shipped to some extent, although with small profit, because of freight charges and the low price of the product.

Of building-stones there is a great variety. Fine granite, both of red and grey colours, is found at many places in the Eastern Townships, and is extensively worked in Stanstead county. Marbles occur in the crystalline series of the same district, and also as a part of the Archæan of the Ottawa area ; while the limestones of the Trenton, Black River and Chazy formations are extensively quarried at many places for building stones, as well as for the manufacture of lime and cement.

Extensive slate quarries are found in eastern Quebec at Melbourne and Danville. Deposits of peat of excellent quality occur in the area along the St. Lawrence above Montreal, and also along the line of the Canadian Pacific Railway east of Montreal and near the Grand Trunk Railway east of Lake St. Peter.

Mineral springs are abundant in Quebec, and large

quantities of mineral waters, such as the Radnor and St. Leon, are bottled.

Petroleum has long been known to exist in the Gaspé peninsula, and extensive boring operations have been carried on in that district during the last ten years to develop, if possible, supplies of mineral oil of commercial value ; but the results of the recent work have not yet been publicly announced.

ONTARIO.

The chief mineral products of Ontario in the year 1899 are enumerated in the following table :—

	QUANTITY.	VALUE.
		$
Copper, lbs................................	5,723,324	1,007,877
Gold, oz...................................	20,340	420,444
Iron Ore, tons............................	25,126	100,806
Nickel, lbs...............................	5,744,000	2,067,840
Silver, oz................................	104,069	62,004
Graphite, tons...........................	1,220	16,179
Natural gas..............................	....	387,098
Petroleum, bbls.........................	808,570	1,202,020
Salt, tons	57,095	234,520
Cement, bbls............................	357,437	550,995
Structural materials and clay products (estimated)....................	....	3,500,000

It will be seen that the structural material class, with petroleum, constitutes nearly fifty-five per cent. of the whole production of the province. Iron and nickel are also important contributors, but none of the other items amount to more than five per cent. of the whole. There is every probability in the future, however, that the metallic mineral products will assume much larger proportions, there having been a decided movement in ate years towards the development of metalliferous mining.

The land area of this province is computed at about 220,000 square miles, making it roughly equivalent to any one of the larger European countries and approximately twice as large as either Great Britain or Italy. Of this large area the southern and eastern portions are most thickly populated, and being underlain by sedimentary fossiliferous rocks of Palæozoic age, and the soils being fertile, agriculture forms the most prominent pursuit.   To this settled region practically all products of the structural material class, as well as the petroleum, natural gas, salt and gypsum are to be credited.   The remaining five-sixths of the province can be best described as a large area of gneissic and granitic Laurentian rocks, throughout which are areas of schists, diorites, etc., of the Huronian system, constituting the chief metallic mineral bearing districts.   In these are found gold-bearing veins and deposits of copper sulphides and nickeliferous pyrrhotite.   In them also occur nearly all the numerous deposits of iron ores, many of which in the east have been more or less worked.   This portion of the province is mostly rugged and rocky, only minor tracts being fitted for agriculture, so that, apart from its forest wealth, mining is likely to constitute its most important industry.

Besides the Huronian areas that have already received a fair measure of attention from prospectors, there are many whose existence has been proved by the work of the Geological Survey, but which, being yet remote from the means of access either by water or rail, have been subjected to little or no exploration for mineral deposits.   These will, no doubt, eventually greatly extend the already great area characterized by the presence of valuable ores.

Besides the Huronian rocks and the fossiliferous Palæozoic strata already alluded to, rocks referred to the lower part of the Cambrian occur in considerable areas on the borders of Lake Superior.   Some of these are equivalent to the Keweenawan series in which are the famous native-copper mines of the United States.   In Ontario such rocks are represented at several points on the eastern shore of the lake and constitute Michipicoten Island as well as St. Ignace Islands in Nipigon Bay; prob- ably agregating about 350 square miles in area. Below this formation lies the Animikie Series, about 700 square miles of which occurs in the vicinity of Thunder Bay, near the west end of the lake. In this series are the veins which have yielded such rich and interesting ores of silver.

The promise of this large northern area of the pro- vince in the way of economic mineral deposits has long been recognized in a general way, and efforts more or less persistent and successful have been made from time to time towardsthe realization of its value. Considerable quantities of silver ores and copper ores, in particular, have been extracted many years ago from the deposits of the Lake Superior region.

In the past, however, none but the most favourably situated and richest deposits could be worked, and the new era in the investigation of the more remote mineral areas of northern Ontario may be said to have commenced with the completion of the Canadian Pacific Railway in 1886. The main line of this system, which traverses these areas for over one thousand miles, with its branch from Sudbury to Sault Ste. Marie and other connected means of communication, have been important factors in the revival and enlargement of the mineral industries of the province.

The gold mining of Ontario is at present centred in Hastings county in the east and in the district lying between Port Arthur and the Manitoba boundary on the west. In the latter district occur the Lake of the Woods, Rainy Lake, Seine River, Shoal Lake, Lake Minnietakie, Wabigoon Lake, and Manitou Lake mining districts. Besides the mines, already developed to a greater or less degree in these districts, prospecting and prelimin- ary work have been carried out at other points in the province, notably on veins in the Huronian areas along the north shore of Lakes Superior and Huron, the numerous discoveries of gold-bearing veins in the Michi- picoten district at the north-eastern corner of the former lake being the latest to come prominently before the public eye.

In these western districts nearly all the gold-ores are free-milling, but in the Hastings county areas, while the gold is in some cases free, it generally occurs in close association with the pyritous and arsenical minerals.

The veins as a rule are of moderate width, although some very large ore-bodies have now been opened up and more or less developed. Early in 1899 several hundred claims were being developed. Of these five were reported as paying dividends, about twenty-five were equipped more or less fully with machinery, and the rest were merely doing prospecting work on a small scale. There were stamp-mills aggregating 300 head of stamps, with a probable addition of 200 more in immediate prospect. About 58,000 tons of ore were milled, producing some 20,000 ounces of gold valued at about $420,000, or an average for the whole province of nearly $7.25 per ton of ore.

A few of the gold mines have attained a depth of from 300 to 500 feet, but most of them are yet quite shallow and not well out of the prospect stage as judged by the standards of older mining districts.

The existence of gold-bearing veins in different parts of Ontario has been known from very early times, and as far back as thirty years ago, or more, they received some attention, especially in the districts more easily reached from the settled portions of the province. Thus the mines of Hastings county caused considerable excitement in the early seventies, and one mine, the Gatling, was worked on a large scale for some years. Of the others in this district, and of those more remote in Lake of the Woods, etc., it cannot be said that they ever got beyond the prospect stage until the revival of mining, which began only a few years ago. The difficulties encountered in the early times having been largely removed by the development of the country, as well as by the improvements in methods of mining, milling and smelting, and gold mining in Ontario is now fast becoming an assured industry.

Silver mining is at present represented by the work in progress at one or two points in the Thunder Bay district west of Port Arthur, consisting in the re-opening of some of the old mines. From 1870 to 1884 this district was rendered famous by reason of the operations at the Silver Islet mine. At this point a vein was found to outcrop on a small rocky islet, about a mile off the mainland, and the contained ore was so exceedingly rich in native silver and argentite that world-wide interest was excited in it. Ore to the value of about $3,250,000 was extracted during the existence of the mine, and the vein was followed to a depth of nearly 1200 feet. The amount of ore in the lower part of the mine being much less than that in the upper, and a number of financial and other difficulties being encountered, it was closed in 1884. Between this point, at the eastern end of the district, and the Silver Mountain mines, some sixty miles to the west, similar veins have been opened at over a dozen points, often to a considerable extent; and in some cases, notably at the Beaver mine, much rich ore has been extracted. Several mills for the extraction of the metal have been erected and worked.

The formation in this district consists of a series of carbonaceous shales and argillites with cherty rocks below them, the whole cut by many dykes and intruded sheets of diabase. Intersecting these rocks are the veins worked, in which native silver and argentite occur either alone or associated with more or less galena and zinc-blende in a gangue of calcite and quartz (often amethystine) with barite and green and purple fluorite. The whole formation lies in an almost horizontal position on the smooth worn surface of the Laurentian gneisses.

Apart from these silver veins proper, deposits of galena have been opened at various points and at different times in Ontario, and a smelter was in operation many years ago at Kingston, working the ore from the Frontenac mine, in Frontenac county. None of these efforts have, however, so far got beyond the trial stages. The silver contents of these ores are generally comparatively low, so that they should be considered rather as lead than as silver ores.

The nickel-copper ores of the Sudbury district have for many years been the sole source of the output of these metals in the province. Previous, however, to the inception of mining at this point, extensive operations were carried on at the mines of the West Canada Copper Co., about 30 miles east of Sault Ste. Marie, on the north shore of Lake Huron. At this group of mines, known respectively as the Bruce, Wellington and Copper Bay properties, the ore consisted of sulphides of copper contained in several large fissure veins cutting the Huronian rock of the vicinity. The underground workings were very extensive and the mines were equipped with a large mill for the extraction of the ore, which was shipped to smelters in England. Owing to the decline in the price of copper, combined with other circumstances, the mines were closed in 1876. The improvement in communications and in the general economic conditions has, however, been so great in late years that renewed interest has now been aroused and an effort is at present being made to re-open these mines and place them once more on a paying basis. The existence of a number of other deposits of copper-sulphide ores in the province was known from early times, but with the above exceptions, none have been extensively worked. Recent and very promising discoveries of copper ores have taken place in the Parry Sound district, near the east coast of Georgian Bay, Lake Huron.

At various times considerable attention has been given to the occurrences of rocks holding native copper, similar to those so well known on the south shore of Lake Superior, of which detached areas are found fringing the Canadian shores of the lake. At Mamainse Point, sixty miles north of Sault Ste. Marie, and at Michipicoten Island, a good deal of money has been expended by three different companies and large plants were erected None of these have been worked for some years now, and the industry can never be said to have reached a really permanent basis. These failures of pioneer effort are, however, so often the experience of the earliest attempts, and have so frequently been followed by ultimate success, that the last word can hardly be considered as said in the matter ; especially as even yet a large proportion of the known areas of the copper-bearing rocks have been but little explored.

The commencement of mining on a large scale at Sudbury may be said to have been when the Canadian Copper Company first began work, in 1886. This company has been in continuous operation ever since, the other operators being the H. H. Vivian Company, the Dominion Copper Company and the Trill Mining Company. The three last named, however, after working for various terms of a few years, closed down, and until quite recently the first-mentioned alone remained in work. In all four cases plants have been erected for smelting the ores, the process adopted being generally similar, viz. : heap-roasting the ores, matting in blast-furnaces and sometimes further enriching the matte by bessemerizing.

The ore consists of a mixture of chalcopyrite and pyrrhotite, occurring in irregular but extensive deposits, in connection with large masses of diabase, intrusive through rocks of the Huronian system. The ore runs from about $1\frac{3}{4}$ to 4 per cent. in copper and about $1\frac{1}{2}$ to $4\frac{1}{2}$ per cent. in nickel. The product shipped consists of a matte averaging about 25 per cent. copper and 18 per cent. nickel, or, where bessemerized, about 40 per cent. copper and 40 per cent. nickel. This matte also carries small percentages of cobalt, platinum, palladium, etc.

About half the nickel supply of the world comes from the Sudbury district and at present practically all the matte is shipped to the United States, where the final stages of the processes of extraction and refining are carried on.

Ores of iron are widely distributed throughout Ontario. In the east, in the counties of Renfrew, Frontenac, Lanark, Hastings, Peterborough and Haliburton, numerous deposits of magnetite and hæmatite have been discovered, and many of them have been worked. These districts are served by the Kingston and Pembroke, the Central Ontario and the Irondale, Bancroft and Ottawa railways. Along the former, a number of the deposits have been opened and much ore has been shipped. These ores are all magnetite, with the exception of the Playfair mine, which was worked many years ago on a body of hæmatite. Along the lines of the two last-mentioned railways, besides bodies of magnetite, several large deposits of hæmatite have been worked in the past. In eastern Ontario little or nothing has been done in the way of iron ore mining, for a number of years the high duty imposed upon ore entering the United States preventing access to the chief market. With the increased demand for ores and the local market afforded by the erection of the Hamilton smelter, several mines have, however, now been reopened.

In western Ontario, along the northern shores of lakes Huron and Superior, a number of deposits of hæmatite and magnetite are known to occur, some of which were discovered many years ago. West of Thunder Bay also, important bodies of similar ores have been taken up. At none of these points, however, has any work been permanently carried on, but much attention is now being directed to them, and important mining operations are about to be begun at Michipicoten.

As early as the year 1800 attempts were made to smelt the iron ores of Ontario, and between that date and 1883 several enterprises were started, only one of which was successful. This one was the furnace at Normandale, in Norfolk county, where the bog-ores of the vicinity were smelted, with charcoal as fuel, and the pig produced was cast into stoves and various useful articles to supply the needs of the early settlers, realizing a profit for the manufacturer. Attempts were also made to smelt the magnetite ores, but these were unremunerative. The fact is that the conditions in a new and sparsely settled country, with very poor communications, were not sufficiently favourable except under special local circumstances.

Now, however, a new departure has been made and

at Hamilton, Deseronto and Midland, modern plants have been recently erected. At the first-mentioned place, only about one quarter of the ore used is Canadian, it being found better to use this ore in admixture with ore from the United States, whence the fuel also comes. These later attempts have of course been much assisted by the protection afforded by the duty on iron entering Canada and the encouragement given by both Provincial and Dominion governments in the way of bounties.

The production of iron ore in Ontario in 1899 was about 25,000 tons, of which practically all was smelted in Canada, a small proportion finding a market in the United States. The amount of pig-iron produced was about 64,750 tons, valued at over $808,000.

The production of petroleum in Ontario dates back to 1860, when oil was first discovered. The oil-producing region is in Lambton county, where the two chief centres are at Petrolia and Oil Springs. The oil-bearing strata are reached by drilling holes to a depth of about 300 to 500 feet until the Corniferous limestone (a member of the Devonian system) is encountered. At first, flowing wells were struck, but all the oil is now obtained by pumping. Thousands of these wells have been bored within a comparatively limited circuit around each of the places above mentioned, the proved oil-bearing areas comprising twenty-four square miles at Petrolia, and one and a quarter square miles at Oil Springs. Within the last year or so, other pools have been receiving attention, notably those at Euphemia, London Road and Dawn, in Lambton Co., and at Bothwell in the adjacent County of Kent. In all these districts the working wells in 1899 numbered about 9000.

The existence of Natural Gas in Ontario was first discovered in 1889, since which date a great many wells have been sunk. The result has been the definition of two fields, one in Essex County and one in Welland County. The greater part of the gas, is piped to the United States, chiefly to the adjacent cities of Buffalo and Detroit. The product is valued at about $300,000 to

$400,000 per annum, the price put upon it being probably rather less than its real value.

Gypsum has been worked for many years along the Grand River, which traverses Brant and Haldimand counties in the peninsula of Ontario. The deposits which are interbedded lenticular masses three to six feet in thickness, occur in the lower portion of the Onondaga formation of the Devonian, and as they lie in a more or less nearly horizontal attitude and crop out on the river bank, they can be easily mined by drifting in from the valley. The value of the crude mineral produced was, for 1899, about $4,000, the output having decreased in late years to about one-third its former amount. The value of the finished products, including plaster, alabastine, etc., amounted to nearly $15,000. Much of the crude gypsum is used locally as a land dressing.

Graphite is mined at one place in Ontario, in the township of Brougham, Renfrew county. The deposit is large, having a thickness of about ten feet, and has been traced for several hundred feet. It consists chiefly of the amorphous variety and considerable shipments are now being made to the United States.

Mica is found in the Archæan rocks of the eastern part of the province and is mined at a number of points. It is chiefly phlogopite and finds a ready market in the United States and elsewhere for electrical purposes. In 1898 the production was about 35 tons, valued at about $8,000. The mode of occurrence and character of the mica is practically identical with that of the adjacent deposits in the Province of Quebec, to which allusion has already been made, and which have been developed on a much larger scale.

Phosphate or apatite is also found and has been worked in Ontario, but as already explained, the prices at present ruling are very unfavourable.

The existence of corundum in Ontario has been among the discoveries of the past few years. It occurs in syenite, and nepheline-syenite rocks which form a belt about seventy-five miles in length, running irregularly from Peterborough county, through Haliburton, Hastings and Renfrew counties. The corundum deposits have now been pretty thoroughly explored in a preliminary way. A strong company has been formed to work some of them and the erection of the necessary mill and plant for the crushing and concentration of the ore is now in progress.

Ontario abounds in building stones of many kinds and often of excellent quality. The old crystalline rocks of the Laurentian country yield granites and gneisses, generally of red or reddish colours, as well as marbles like those of Arnprior and Barrie. Limestones and sandstones are quarried in a great number of places in the southern and thickly inhabited parts of the province, chiefly for local use, but also for the supply of the larger cities and to a small extent for export. Clays and shales of different kinds are largely employed in making bricks, drain-tiles, terra-cotta, etc. The manufacture of lime and hydraulic cement also constitute important industries, deposits of shell-marl being utilized to a considerable extent for the last-named purpose. It will be observed that, taken together, materials applicable to purposes of construction represent a large proportion of the total mineral output of Ontario.

MANITOBA.

This is an agricultural province, situated almost in the geographical centre of Canada as between the Atlantic and Pacific ; but as it overlaps on the east the region of crystalline rocks in which the gold mines of western Ontario are situated, it is not improbable that mines of the same kind will be developed locally within its limits. Iron ore has already been discovered on an island in Lake Winnipeg ; gypsum deposits are known near Lake Winnipegosis ; and salt has been made from brine springs on the west side of Lake Manitoba. Dolomitic limestone, excellent for building purposes, is quarried in several places and employed in construction in Winnipeg, and brick clays of good quality are abundant.

THE NORTH-WEST TERRITORY.

With the exception of the Yukon district, to which special notice is given further on, the several districts comprised in the North-West Territory require but brief mention here. The southern part of this great region, between Manitoba and the Rocky Mountains, is pre-eminently an agricultural and pastoral country, under-lain by unaltered rocks, chiefly of Cretaceous age and generally not productive of economic minerals except coal, building stones and clays suitable for brick-making and the manufacture of some kinds of pottery. The Mackenzie district, the district of Keewatin and the Franklin district, the latter comprising the great Arctic archipelago to the north of the continent, contain large areas of the older rocks in which metalliferous and other minerals will no doubt eventually be developed, but in regard to which little is yet known, being at present too remote for practical exploitation.

The mineral fuels of the country west of Manitoba possess great prospective importance and considerable present value. Without any very great differences in regard to geological age, they are found to improve in character as the Rocky Mountains are approached. The Souris River country and the region about Medicine Hat, in Assiniboia, yield lignite only. In western Alberta excellent fuels which may be described as lignite-coals occur, and are already somewhat extensively worked at Lethbridge. In the foot-hills adjacent to the mountains, are many deposits of bituminous coal, of which some are worked to a limited extent. On the Bow River valley, within the outer range of the Rocky Mountains, steam-coal and anthracite are produced at Canmore and near Banff. The output is limited in each case only by the requirements of the available market, for the productive capacity of the deposits in the aggregate is enormous. The Crow's Nest coal-field, also in the Rocky Mountains, but situated to the west of the watershed, is referred to in connection with British Columbia.

The lignites of the eastern plains, although not so

valuable as true coals, pessess great local importance as sources of supply of fuel for adjacent farming communities in the prairie country. In part of the Souris country it has been estimated that the lignite underlying each square mile of surface amounts to over 7,000,000 tons. Farther west, the easily available fuel contained in the seam worked at Lethbridge, in its course between the Belly and Bow rivers, has been estimated at 330,000,000 tons. The coal-bearing region of the North-West Territory between the International boundary and the 56th degree of latitude is approximately 65,000 square miles in area.

Lignites and lignite-coals are also known in the Mackenzie district and these may soon become useful in connection with the navigation of the Mackenzie River, upon which steamers already ply. Coals referable to the Carboniferous period proper occur in the Parry Islands in the extreme north, and fuels of Cretaceous or Tertiary age in Grinnell L-: l of the Franklin district; but for all present purposes ɩ..cse are inaccessible.

Along ɩhe Athabasca River, in the district of the same name, the lower sandstones of the Cretaceous formation, where they come to the surface, are saturated with great quantities of bitumen or maltha. They have absorbed this from the underlying Devonian rocks, from which it has no doubt originally flowed out in the form of petroleum. There is every evidence of the existence of a great petroleum field in the northern part of Alberta, in Atha basca and perhaps even further north. Some experimental borings have already been carried out by the Geological Survey in this region, but so far without reaching beds yielding petroleum of commercial value. Natural gas has, however, been found in large quantity in two of these borings and in several borings further to the south in Alberta, and there is undoubtedly a large area of the North-West where this convenient fuel may be obtained without difficulty when required.

Gold has long been worked on a limited scale along the western part of the North Saskatchewan River, above and below Edmonton. In late years steam dredges have been placed on the river and it is probable that their operation may soon result in the development of a considerable local gold production. The gold found is all very fine and evidently far-travelled. It characterizes the upper parts of many other streams in western Alberta and Athabasca, including the Peace River.

Clay ironstone occurs in many parts of the North-West, but the manufacture of iron is not to be looked for in the near future in this region. Copper ores occur in the eastern part of the Rocky Mountains included in Alberta and zinc-blende has been found.

Salt, in the form of brine springs issuing from the Devonian rocks and accumulating in crystals by natural evaporation, occurs near the Mackenzie River in the northern part of Alberta, and gypsum is found on the Peace River.

In the far north, the copper, lead and possibly gold ores of the vicinity of Great Slave Lake ; the native copper deposits of the Coppermine River and the copper ores of the north-west shores of Hudson Bay, will some day, no doubt, be utilized ; but no rapid development of mineral resources in these regions need be looked for, except in the possible event of the discovery of gold-placers like those of the Yukon which might result in the forced establishment of means of communication with the outer world.

BRITISH COLUMBIA.

The province of British Columbia, includes the mountainous western country of Canada as far north as the 60th degree of latitude. This is a portion of the "Cordilleran belt," resembling the Pacific borders of South America, Mexico and the United States, and like these characterized throughout its length by rich metal-liferous deposits. The area of British Columbia is about 383,000 square miles, or greater than that of any country in Europe except Russia.

Its development, as a mining country, has been rapid during the last few years, but can only be said to have begun, systematic work being so far confined to a few districts of comparatively limited size. If we include with British Columbia the adjacent and similarly characterized Yukon district to the north, we comprise what must be regarded as pre-eminently the great mining region of Canada, and one which will in the near future be recognized as one of the most important in the world.

Ten years after the great influx of gold miners to California, a similar and scarcely less precipitate movement occurred toward British Columbia, in consequence of the discovery of rich placer-gold deposits on the Fraser and its tributaries. The gold was traced up to Cariboo, far inland; population followed, and in 1863 the output of gold was valued at nearly four million dollars. This was its maximum year, and thereafter it slowly declined as the easily worked and very rich deposits became more or less exhausted. New fields, such as Big Bend, Ominica and Cassiar were subsequently discovered, but none of them equalled the Cariboo placers in value. It remains to be seen whether Atlin, the latest gold-field, will rival Cariboo. Coal had, many years previously, been found and worked in a small way on Vancouver Island, and the output of coal continued gradually to increase, but the absence of suitable means of communication in the interior of the province, for many years prevented the exploitation of the lodes from which the alluvial deposits of gold had been derived, as well as that of other metalliferous deposits. In 1893, the output of gold, all from placer-deposits, had dwindled to less than $400,000. The construction of the Canadian Pacific Railway through the southern part of the province, soon gave, however, the required impetus to the development of its mining resources. About 1886, when trains first began to run regularly between the Atlantic and Pacific coasts of Canada, valuable discoveries of ores were made in the West Kootenay district. These were promptly followed up by branch lines of railway and by steamer service on the lakes and rivers of that part of the country.' The East Kootenay district, various parts of the Yale district and several regions on the coast also came into prominence; and, generally speaking, wherever adequate means of transportation have been provided, mining enterprises of importance have grown up on the steps of the prospectors.

Preliminary geological investigations have already been extended over the greater part of British Columbia, and their result is such as to indicate that mineral deposits of equal value to those of which the working has already begun, will be recognised in many new districts as these become accessible to the miner. At the present time, prospecting and discovery are far in advance of real mining requiring capital and fixed conditions. The number of established and remunerative mines is comparatively small, that of claims and locations of promise is very great. Most of these may probably never develope into active mines, but if even one per cent. should so develope, British Columbia will be a hive of mining industry within a few years. In 1899, over 11,000 new mineral claims were recorded, while up to date the Crown grants, obtained in British Columbia, after the expenditure called for by law, number about 2,000.

It is impossible in an article such as this to follow the interesting history of the discovery and development of minerals and mining in British Columbia in any detail. It will be endeavoured only, and in a few words, to sum up the existing conditions. The actual amount and value of the mineral products are given in the subjoined table, derived, together with many other facts, from the official reports of Mr. W. F. Robertson, the Provincial Mineralogist.

Production of Minerals in British Columbia.

		1898		1899	
		Quantity.	Value.	Quantity.	Value.
Gold, placer. ...Ounces		32,167	$ 643,346	67,245	$ 1,344,900
" lode ..... "		110,061	2,201,217	138,315	2,857,573
Silver .... "		4,292,401	2,375,841	2,939,413	1,663,708
Copper............Pounds............		7,271,678	874,784	7,722,591	1,351,453
Lead............ "		31,693,559	1,077,581	21,862,436	878,870
Coal..... ....Tons, 2,240 lbs..		1,135,865	3,407,595	1,306,324	3,918,972
Coke..... " "		35,000	175,000	34,251	171,255
Other Materials.......		....	151,500	....	206,400
			$10,906,861		$12,393,131

It is necessary to include the past two years in the above statement, because certain local difficulties connected with labour, and in no wise dependent on the mines themselves, have reduced or stopped the output in certain cases during the year 1899. The general increase for the year is also, in consequence, much less than it otherwise would have been. As compared with that of 1898 it is equivalent to over 13 per cent. Mr. Robertson estimates that but for the temporary shutting down of several mines it would have amounted to 27 per cent.

In 1890, the mineral products of the province aggregated $2,608,803 in value. They are valued in 1899 at $12,393,131. The trend of mining development is well indicated by the fact that the products of lode-mining are now more than twenty times greater than they were in 1893.

The table given above broadly indicates the nature of the leading mineral products of British Columbia, but it may be of interest, following its order, to add a few notes briefly characterizing the several classes of products.

The output of gold is now greater than ever before, considerably surpassing that of 1863, which marked the climax of placer-mining. Placer or alluvial mining is still important, but hydraulic methods, involving large capital for their initiation, but then dealing easily and at small cost with great quantities of comparatively low grade gravels, have now in great part superseded the more primitive modes of work. This is especially the case in the Cariboo mining district, where large sums are being spent in constructing water-ditches of great length and in installing expensive machinery. Attention has also been lately given to river-dredging for gold on the Fraser and Quesnel, with promising, but as yet not very important results. The recent rich discoveries in Atlin district, in the extreme north of the province, have locally revived old methods of placer-mining, but the product of lode-mining is now much greater than that from the placers. A certain number of mines are in operation, chiefly in **the Bound**ary and Okanagan districts, of which the ore is treated in the stamp-mill, yielding free gold and rich concentrates; but the greater part of the yield comes from the smelting ores of Rossland, supplemented by those of the Ymir mines of the Nelson district. These ores consist **chiefly** of pyrrhotite and copper-pyrites, from which a rich auriferous copper-matte, or refined copper carrying the gold, is produced. The values, by smelter returns for Rossland ores, vary from about $10 to about $30. In 1899 the output of ore at Rossland was 172,665 tons.

Silver is obtained chiefly from the argentiferous galena ores of the Slocan and Ainsworth divisions of West Kootenay and from similar ores in East Kootenay. There is, however, another important class of ores containing silver and copper, of which the Hall mines in the vicinity of Nelson, West Kootenay, are typical. Although the output of silver has been affected by the low price of the metal, as well as especially by the closing-down of a number of mines above alluded to, the deposits already developed are most important, and, because of the high percentage of silver they contain, are certain to become very large producers even at present prices, when their working is resumed. The silver-lead ore raised in West Kootenay in 1898 amounted to 32,000 tons, containing on the average about 97 ounces of silver to the ton and 47 per cent. of lead. The output of the Hall mines in the same year amounted to 45,000 tons, containing from 15 to 20 ounces of silver per ton and 2 to $2\frac{1}{2}$ per cent. of copper.

Argentiferous galena ores are, in addition, known in British Columbia as well as in the Yukon district, in many places where they at present possess practically no value, but lie in reserve, awaiting the future progress of railway construction.

The copper production of British Columbia is, as yet, only in its initial stages, **and,** if present prices continue will without doubt soon greatly increase. It is now chiefly the result of treatment of the gold-copper ores of Rossland camp and the silver-copper ores of Nelson. Some copper is, however, also being produced from mines in Vancouver Island and along the mainland coast of the province. In the Boundary district, adjacent to Ross-land, very large deposits of low-grade copper ores exist upon which development is now progressing. Copper is also known to occur in many other parts of the province, but it can be utilized only where means of transportation have be provided.

From what has already been stated, it will be understood that lead, in the form of argentiferous galena, is closely associated with the silver product. These two metals must be worked together and conditions affecting one also affect the other. The output of lead is already very considerable and it is certain largely to increase.

The coal output of 1899 was greater than that of any previous year, by far the larger part being derived from the mines of Nanaimo and Comox in Vancouver Island. Most of this coal is exported to California, but shipments are also made to Alaska and to many other Pacific ports. The very remarkable coal-field of the Crow's Nest Pass, in the Rocky Mountains, is now, however, also beginning to produce a considerable amount of coal and is already supplying the greater part of the coke employed in smelting in the West Kootenay district.

Coal mining is at present confined to certain points on the coast, where it has become an established industry and to the Crow's Nest field just alluded to ; but coal, lignite-coal and lignite are known to occur in many other parts of the province. In the Queen Charlotte Islands anthracite is found. It is not possible here even to name the various localities in which mineral fuels of different classes occur, possessing at least a local or a prospective importance. It is interesting to observe that all these fuels are referable either to the Cretaceous or the Tertiary periods, very much later in date of origin than the coals of Eastern Canada or those of Great Britain. In regard to quality, many of them are excellent.

A total amount of $206,400 is made up in the returns of 1899 by other minerals of various kinds, of which no one is at present of considerable importance, but several are of interest and may eventually become of great value. Platinum, found in association with gold, particularly on the Similkameen River, is one of these, and cinnabar, discovered in the vicinity of Kamloops Lake is another. The ores and minerals already known to occur in British Columbia are so numerous and varied that no attempt is made to enumerate them here. Reference has been made only to those which already, in the course of discovery and development, have achieved greater or less prominence.

YUKON DISTRICT.

To the north of the province of British Columbia, in the extreme north-west of Canada, lying between the Mackenzie River and the United States territory known as Alaska, is the Yukon district. It is for the most part drained by streams tributary to the great river from which its name is taken. Until very lately, it has been one of the most remote and least frequented regions of the world, inhabited by a sparse native population and yielding only some small product in skins and furs. Posts of the Hudson's Bay Company had been established in the district many years ago, but for the most part soon abandoned. It was generally regarded as an arctic solitude, although even the facts long ago recorded might have contradicted any such belief, for the summer is sufficiently long for the growth of crops almost to the Arctic circle and the winter is not more severe than that of Manitoba.

About 1878, miners began to enter this region, and gold mining may be said to have begun on a small scale on the river-bars of the Lewes and Salmon in 1881 and 1882, on the Stewart in 1882 or 1883. In 1886, late in the autumn, "coarse" gold was found for the first time on Forty-mile River, a tributary joining the Yukon from the west near the Alaskan boundary. The few hundred miners then in the district concentrated at Forty-mile in 1887, and, following up its tributaries, found rich ground. Thence, the productive field was gradually extended southward across the local watershed to the tributaries of

Sixty-mile River. Miners came in in greater numbers every year, so that, when in the autumn of 1896 the very rich discoveries on the Klondike River became known, there was a considerable population ready to take advantage of them. A wholesale migration or "stampede" then occurred from the older mining centres to the new find, and in 1897 and 1898 a probably unexampled rush of fortune-seekers from all parts of the world took place. Great hardships were necessarily endured by many of these people, who, without experience or proper appliances endeavoured to force their way to this new Eldorado. A few of them realized fortunes, many returned broken and disappointed, but the phenomenally rich character of the placers of the Klondike region was made apparent by the work done in it, and a busy town was established as its centre.

Greatly improved means of access have since been established. A short railway has been built from the sea to the head of Lewes River and a number of fine steamers have been placed upon that river and on the Yukon. A telegraph line has been built by the Canadian Government from Skagway, on the coast, to Dawson, the capital of the Klondike region, and the construction of an overland line to connect this with the lines of British Columbia and the telegraphic system of the world is now in progress.

Such is a very brief outline of the events leading up to the present conditions in the Klondike region—the latest great gold discovery. Other similar regions in this wide Yukon district may remain to be found, but in the Klondike, matters are now established on a substantial and orderly basis, and during the past three years about $28,500,000 worth of gold has been produced, the year 1899 being credited with $16,000,000 with the prospect of a largely increased output in 1900.

The Klondike gold-fields are situated in a tract of country of some 800 square miles in area between Klondike and Indian rivers, affluents of the Yukon River near the 64th degree of north latitude. The region may be discribed as a high plateau, deeply trenched by the wide flat-bottomed valleys of a number of tributaries of the rivers just mentioned. These rise together near the central part of the original plateau, which constitutes the highest ground in the vicinity and is known as The Dome. Very numerous short and narrow tributary valleys and gulches join the several larger streams along their courses, to nearly all of which local names have now been applied.

The gold is found and worked in the gravel deposits of the valleys and their adjacent slopes. It is evidently local in its origin, for it is usually but little worn by attrition and often still contains quartz, and the associated gravels themselves are composed solely of the rocks of the immediate vicinity. So far, little gold has been discovered in the parent rock, but lode-mining or "quartz mining" may confidently be looked forward to in the near future in such a district.

Several classes of gravel deposits are recognized and have been described by Mr. R. G. McConnell of the Geological Survey. The most important of these are those known as the "quartz drift" and the "stream gravels." The first occur at some height above the present streams, covering the nearly flat rocky floors of older and shallower valleys, into which the streams have, at a later period, cut the deep valleys in which they now run. The quartz drift thus represents an ancient auriferous deposit washed down from the decaying rocks of the neighbouring plateau, while the stream gravels contain the further-concentrated gold derived from portions of the quartz drift which have been re-washed, together with additional gold resulting from the later excavation of the streams in the rocks of the district.

It is perhaps difficult to make the origin of these deposits clear by a description so brief as can be accorded to them here, but the result of the long natural processes of wearing down and concentration has been to accumulate a great body of generally rich gold-bearing gravels in and about these valleys of the Klondike. It seems probable, in fact, from the examinations already made, that gold to the value of at least $95,000,000 will, within the next few years, be produced from the gravel deposits ·

of this region.   The aggregate length of the payable portions of the valleys already known, is about fifty miles, of
which some parts are phenomenally rich, as several claims
only 500 feet in length along the streams will each produce, when fully worked, more than a millon dollars
worth of gold.   Many claims will produce half a million
and a great many are valued at between half and a quarter of a million.

The principal gold-bearing streams of the Klondike,
are : Bonanza Creek, with its tributary Eldorado Creek ;
Bear Creek ; Hunker Creek with its tributaries Quartz
and Dominion creeks ; Gold Run Creek and Sulphur
Creek.

Attention has been almost entirely directed to the
Klondike region since its discovery, to the neglect of the
streams first worked in the Yukon district, and somewhat
to the detriment of general prospecting, which may be
expected to result in further discoveries of the same kind.
A mining industry has now, however, become established
in the Yukon district which will ultimately lead to the
exploitation of its mineral wealth in all its extent and in
various lines.   Good lignite-coal is known to exist and is
already worked to a small extent in several places, copper
has been discovered, both in the form of the native
metal and as sulphide ores, argentiferous galena has been
found and gold-bearing quartz in payable deposits is
almost certain to be developed.   The future of the Yukon
district as a mining region seems, therefore, to be
assured, in conformity with the general forecast which
has long ago been made in regard to the prospective
mineral wealth of each and every considerable part of
the Western Cordillera.